Silver Bream

This book is dedicated to the silver bream, a fish too long neglected by anglers, wildlife enthusiasts and science.

In fact, this is the first book ever devoted to this freshwater fish.

Scientist, author and broadcaster **Professor Mark Everard** introduces the biology of the silver bream, angling for this fish, and its diverse social quirks and values.

MARK EVERARD

Silver Bream
Britain's Most Neglected Freshwater Fish

CRC Press
Taylor & Francis Group
Boca Raton London

CRC Press is an imprint of the
Taylor & Francis Group, an **informa** business

First edition published 2023
by CRC Press
6000 Broken Sound Parkway NW, Suite 300, Boca Raton, FL 33487–2742

and by CRC Press
4 Park Square, Milton Park, Abingdon, Oxon, OX14 4RN

CRC Press is an imprint of Taylor & Francis Group, LLC

Library of Congress Cataloging-in-Publication Data
A catalog record for this book has been requested

ISBN: 978-1-032-31734-2 (hbk)
ISBN: 978-1-032-31733-5 (pbk)
ISBN: 978-1-003-31104-1 (ebk)

DOI: 10.1201/9781003311041

Typeset in Joanna MT
by Apex CoVantage, LLC

Contents

One

Is the silver bream (Blicca bjoerkna) Britain's most neglected freshwater fish?

I am constantly amazed by the number of anglers, including some otherwise experienced ones, who tell me that they have no idea about how to identify a silver bream. A minority even admit that they had not heard of the silver bream.

Leafing through my not-insubstantial fishy library, I also see that many angling books fail to give any attention to silver bream. Others give it, at best, a token mention, often as an adjunct when dealing with common bream. The internet also generally addresses this fish in bare minimum terms, many web pages replicating the same rather tired text.

Many smaller British freshwater fishes – two species each of loach and stickleback, minnows, bullheads, ruffe, bleak and minnows, and also non-native bitterling, sunbleak and topmouth gudgeon – are patronisingly described as 'minor species'. But even these allegedly lesser species are written about in terms of their curious characteristics, conservation status or roles in food webs.

Not so the silver bream! The silver bream is too big to be classed as a 'minor species'. It is also not now unduly restricted in distribution. Yet the silver bream seems largely to have swum beneath the radar of much of the angling, nature conservation and scientific communities. This neglect is a shame, as the silver bream is a handsome fish, growing to approximately the same size as more widely appreciated 'silver' fishes such as roach and rudd.

DOI: 10.1201/9781003311041-1

The structure of many of the books I have written on fish follows a pattern of getting to know the fish (their biology), moving on to angling considerations and then addressing their cultural associations and context. The poor old silver bream though almost entirely lacks cultural associations!

However, by great good fortune – for the silver bream at least if not me – my affection for this handsome smaller fish far outweighs my meagre grasp of the economics of publishing. I feel that this most neglected of British freshwater fishes must be neglected no more!

This book then is dedicated solely to the silver bream. May all who already know this fish, or will come to know it, be glad that its time for immortalisation in print has finally come!

Two

The silver bream is known by science today by the Latin name *Blicca bjoerkna* (Linnaeus 1758). It is also known by several other common names, including the white bream, the bream flat or breamflat and, formerly, the Pomeranian bream. A wider history of the Latin and common names applied to this fish is covered in Chapter 3 of this book.

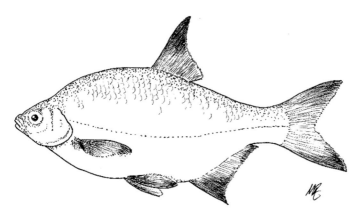

The silver bream is no giant in the world of freshwater fishes. Although individuals have been recorded in excess of 4 lb (1.8 kg) across the broad European and northern Asian range of the species, any fish over 1 lb (0.45 kg) is really quite a specimen, especially in the British Isles. The maximum length recorded in scientific literature is 45.5 centimetres (18 inches), though a length of 20 centimetres (8 inches) is more common.

DOI: 10.1201/9781003311041-2

FEATURES OF THE SILVER BREAM

The silver bream is a deep-bodied fish, notably flattened laterally (from side to side). Like other members of the *Leuscicidae* (the minnow family), the body of the silver bream is evenly covered in regular scales. These scales are large, almost equal to the diameter of the eye. (By contrast, the scales of the common bream, *Abramis brama*, are considerably denser and smaller.)

The scales of silver bream are covered by a layer of mucous, as is common in related fish, though it is a misnomer that silver bream are as 'slimy' as common bream. In reality, silver bream have considerably less loose mucus than common bream, with assertions that they are similar in this regard most likely based on lack of familiarity or unfounded assumptions. One such incorrect description is found in the 1969 *A Ladybird Book about Coarse Fishing*:

> *Both species are generally covered in an exceptionally thick*
> *layer of protective slime. It drips off as they are netted from the*
> *water and covers everything that the fish contacts.*

The common name 'silver bream', along with others such as 'white bream', celebrates the often-extreme silvery reflectiveness of these fish. Though the fish appears brilliantly shiny when held out of water, this colouration is a device to help it merge into its surroundings by reflecting their subtler hues. The underside of the fish bleeds into silvery-whitish, with darker tones on the back that vary with the surroundings in which the fish lives.

In his authoritative 1969 book *The Fishes of the British Isles and North West Europe*, Alwyne Wheeler notes:

> *The head and back are olive brown or greyish, the sides pale*
> *with a silvery sheen, and ventrally it is white. The dorsal, tail*
> *and anal fins are grey tinted, the pelvic fin tip dark, reddish*
> *near the base.*

Along each flank, there is a prominent and complete lateral line. This row of sensory pits extends the whole length of the body from behind the gill cover (operculum) through to the caudal peduncle (the root of the tail). There are 44–48 scales along the lateral line, each punctuated, allowing water to enter the sensory pits by which the bream senses water movement. Within the body, the backbone comprises 39–40 vertebrae, an internal feature that is sometimes useful in distinguishing this species from others.

The dorsal fin is tall but not excessively long, tending to be relatively taller and more pointed than that of the common bream. The dorsal fin is located on the highest point of the back and is supported by three spines and 8 or 9 soft branched rays (III/8–9 in scientific notation). Beneath and to the rear of the body behind the anus, the anal fin is elongated, slightly concave in the outer margin and supported by III/21–23 spines and soft rays. The bases of the paired fins (the pectoral fins behind the gill and the ventral or pelvic fins beneath the body) have a pronounced orange or reddish tint, a feature often useful in distinguishing them from small common bream (though it is not uncommon for the pectoral fins of smaller silver bream to have some pale orange tint). The caudal (or tail) fin is supported by 17–19 soft rays.

The mouth of the silver bream is sub-inferior (slightly downward-inclined), indicating a preference for feeding on the bottom of the waterbodies in which it occurs. Whereas the common bream can extend its mouth as a tube, one of the key external diagnostic features of the silver bream is that the mouth can't be thus extended. In that regard, the mouth of the silver bream is similar to the mouth of fish such as roach (Rutilus rutilus) and rudd (Scardinius erythrophthalmus). Akin to all other members of the true minnow family, Leucisidae (discussed later in this chapter), the mouth of the silver bream lacks barbels (or 'whiskers').

Another of the more obvious features distinguishing silver bream from common bream is the size of the eye. The eye of the silver bream is large, with a diameter about equal to or greater than the length of the snout in adult fish larger than 10 centimetres (4 inches). That of the common bream is substantially smaller.

Bent J. Muus and Preben Dahlstrom record in their 1967 *Collins Guide to the Freshwater Fishes of Britain and Europe*:

> The eye diameter is greater than, or as long as, the length of the snout.

SILVER BREAM DISTRIBUTION

Silver bream are widespread across Europe from southern Sweden to north of the Alps, and across northern Asia westwards to the former USSR. They occur naturally in drainage basins discharging into the North, Baltic, White and Caspian Seas as well as the Black Sea south to the Rioni drainage reaching into western Georgia. They are also distributed across the Aral, Marmara and Anatolian Black Sea basins west of Ankara. They are also found in Atlantic basins southward as far as the Adour drainage in France and the Mediterranean basin in the Hérault and Rhône drainages of France.

Silver bream are naturally absent from the Iberian Peninsula, Italy, the Adriatic basin, Crimea, Scandinavia, north of Sundsvall (Sweden) and 65°N (Finland). The fish has been introduced locally into Spain and north-eastern Italy. The species is also now found in the west of France, south of the Loire, though it is probably introduced there; it has also apparently been introduced into small coastal drainages of Var, a river system that discharges to the Mediterranean at Nice. In the British Isles, the natural range of the silver bream is in eastern-flowing rivers.

In his 1943 book *Coarse Fish*, Eric Marshall-Hardy, one-time editor of *Angling* magazine, wrote:

> As to the distribution of silver or white bream, they are common on the Continent, but may be regarded as rare in

*the British Isles, occurring only in the Eastern Counties from
Suffolk to Yorkshire, and there not in large numbers. . . .
Sluggish rivers and still waters are the natural habitat of both
bronze and silver bream.*

In his 1983 book The Complete Freshwater Fishes of the British Isles,
Jonathan Newdick records:

*The indigenous Silver bream is locally common in parts of
England, particularly the eastern Midlands. It is scarce north of
Yorkshire, rare in Wales and absent from Ireland and Scotland.*

The natural British distribution of silver bream, before we
humans started transferring them and many other fishes across
impenetrable catchment boundaries, is explained by the geological history of the British Isles. Much of the land mass of
mainland Britain was in fact not an island prior to the latter part
of the last Ice Age, between 6,500 and 6,200 BC. Up until this
time, Britain was connected to the European continent by a land
bridge known as Doggerland. To the south of Doggerland, the
Rivers Thames, draining eastwards across what is now Britain,
and the Rhine and Scheldt in contemporary continental Europe
were all part of a far larger Channel River. The combined flows
of these major tributaries of the greater catchment ran south-
wards, discharging into the Atlantic Ocean.

Towards the end of the last Ice Age, a catastrophic
megatsunami occurred as a gigantic ice lake to the north
of Doggerland broke. This massive flow of released water
inundated and then cut through Doggerland. This event
separated what we know today as the British Isles from continental Europe.

The former land bridge remains now only in name as the
Dogger Bank, forming part of the bed of the southern North
Sea. Likewise, the lower part of the Channel River is now the
English Channel, the English and Continental European tribu-
tary rivers separated into their current forms.

This geological history explains the limited natural distribution of many species of freshwater fish in the British Isles. Species such as barbel, gudgeon, burbot, ruffe and many more, including silver bream, were limited to catchments formerly connected to continental European rivers in the greater Channel River catchment. Silver bream and many other native species were originally limited to present-day British catchments from the Thames northwards to the Humber. Terrestrial barriers between catchments prevented these native fish from spreading westward and northward across the land mass.

However, silver bream, amongst other fishes naturally restricted in range, have since been widely distributed across England at the hands of humanity. This includes their spread locally across the south, south-west, midlands and north-west of England.

At the time of writing, silver bream remain absent from the island of Ireland, as also Scotland and Wales. Lacking a historic land and river bridge connection, silver bream are naturally absent from Ireland. Fishes present in the island of Ireland were documented by *Giraldus Cambrensis* ('Gerald of Wales' c.1146–c.1223), a medieval clergyman and chronicler of his times, in his book *The History and Topography of Ireland*. Giraldus Cambrensis recorded that "*. . .pike, perch, roach, gardon, gudgeon, minnow, loach, bullheads and verones. . .*" were absent from Ireland. Of the freshwater fishes recorded by Giraldus as found in Ireland, all are tolerant of salt water as either migratory or brackish water species. These include brown trout, Atlantic salmon and Arctic charr as well as pollan, three-spined sticklebacks, European eels, smelt, shad, three species of lamprey and the (far from common) common sturgeon. These fishes were all able naturally to colonise Ireland's freshwater ecosystems without the aid of humanity due to their salt tolerance. As we know from the many introductions of coarse freshwater species unable to tolerate marine conditions – common bream, rudd, roach, dace, gudgeon and tench amongst others – Ireland's diverse fresh waters are otherwise ideal, many introduced species not only become established but also are spreading, except in the most inhospitably turbulent environments such as spate rivers. Though, as noted, silver bream have not thus far been introduced into Ireland at the time of writing, they would doubtless thrive in the diverse still and slow-flowing, rich lowland waters of Ireland, however at the cost of changing the characteristics of natural ecosystems.

THE SPREAD OF SILVER BREAM

Virtually, all the spread of silver bream from their natural British and wider European and northern Asian distribution can be attributed to the actions of humanity. Whether for food, sport, aquaculture or other deliberate purposes, through canals and other man-made connections, or as stowaways in stockings of other fish species, silver bream have had their range extended.

We know little of the detail of the deliberate spread of silver bream throughout England. However, an interesting historic record is found in the 1987 book *The Bristol Avon: Fish, Freshwater Life and Fishing*, written by D.E. Tucker. Tucker, formerly fisheries manager in the Bristol Avon, wrote that silver bream were naturally absent from the Bristol Avon but that

> . . . in 1903, the Bath Anglers' Association planted a consignment of young silver bream from the eastern counties into the Avon upstream of Bathampton Weir; two were caught about a year later weighing 1¾ lb (0.8 kg) and 3 lb (1.4 kg). These weights are heavy for typical silver bream, and one wonders whether the planted fish were really silver bream of were probably common bream, the young of both species being practically indistinguishable.

There is an urban myth, often repeated and widely accepted as a truth, that the eggs of fish attached to water plants can get transported to new waters entangled in the feet of ducks.

However, no shred of scientific evidence has emerged to support this belief. In fact, ducks are remarkably streamlined and so far from likely to carry tangles of vegetation in flight.

However, a really interesting scientific study was published in 2020 posing some interesting questions. The study found that, when captive mallard ducks were fed with vegetation to which the developing eggs of two species of carp (common carp and Prussian carp) were attached, a proportion of live embryos could be retrieved from their faeces. Even more remarkably, some of these subsequently hatched into viable larvae. This newly demonstrated phenomenon is known as 'endozoochory'. Silver bream are amongst a range of coarse fish species that release their eggs on marginal vegetation much as carp. More research is required to test quite how viable endozoochory is as a realistic mechanism for the spread of hardy carps and less hardy silver bream.

SILVER BREAM HABITATS AND HABITS

Silver bream are fish of slow and still fresh waters. They are found in temperate north-western regions of Eurasia between latitudes 65°N and 40°N, and where temperatures range from 4°C to 20°C.

They may be abundant in suitable habitats, which include the stagnant waters of lakes and reservoirs as well as more gently flowing rivers. In ideal locations, such as eutrophic lakes and dam reservoirs in Turkey, silver bream are known to constitute the main component of the ichthyofauna. Silver bream appear to prosper where they are present in canals, offering still or slow-flowing, enriched water conditions.

Silver bream can also tolerate moderately brackish water. A report by the Baltic Marine Environment Protection Commission of the coastal fishes of the Baltic Sea in 2004 found:

White bream (Blicca bjoerkna), zander, Northern pike, and bleak were caught in considerable numbers in most areas.

Silver bream are gregarious by nature, frequently found in shoals. Once located by the angler, a number of silver bream may be caught in a sitting.

In their 2007 *Handbook of European Freshwater Fishes*, Kottelat and Freyhof state that silver bream are predominantly nocturnal species. This observation may differ from that of many anglers who catch them by day, though, like common bream, there may be pronounced crepuscular (dawn and dusk) periods of most intense feeding. It is certainly my experience that silver bream do not necessarily feed voraciously at night in the summer, often ceasing to feed in full darkness. However, as waters clear in pools and canals in winter, I have found that they do indeed switch to feeding activity in the first hour or two of darkness. The cover of murkier summer water, and that of darkness in clearer winter conditions, may be an important trigger to feeding. Scientific studies have found that silver bream move into the margins of large rivers to forage for food by night.

THE SILVER BREAM DIET

The silver bream is described in a great deal of the scientific literature as having a diet predominantly including invertebrates. These include both planktonic invertebrates (mainly small crustaceans and insects living in suspension in the water column) and molluscs, crustaceans, worms and insect larvae (particularly those of chironomids that are also known as 'bloodworms') on the bed of the river or still water. This overlaps strongly with the dietary preferences of the common bream. As described in a 2017 paper by Okan Yazıcıoğlu and colleagues (see Bibliography):

> This species graze upon aquatic invertebrate (zooplankton, molluscs and chironomids) and penetrate sediment.

However, as many anglers will know, silver bream are partial to bread, sweetcorn, stewed wheat and other vegetable-based baits. In fact, the diet of the silver bream appears to be broad and omnivorous, in all probability varying throughout the year in a similar manner to that of the roach.

In his 1994 book *Freshwater Fish of the British Isles: A Guide for Anglers and Naturalists*, Nick Giles observed the physical differences between the mouth of silver bream and that of some other fishes, and how these can affect diet. Giles observed of (common) bream and carp that "*. . . they suck up mouthfuls of sediment and then blow it back out through their gills*" but conversely silver bream, tench and eels "*. . . also probe the surface of sediments for food, digging much more shallowly than common bream or carp*".

In his 1969 book *The Fishes of the British Isles and North West Europe*, Alwyne Wheeler states:

> *Young fish eat small crustaceans, many of them planktonic and including copepods, cladocerans (Chydorus) and ostracods. They also eat higher plants and green algae. Around maturity they assume a diet composed mainly of bottom-living organisms, for example chironomid larvae (bloodworms), caddis larvae, mayfly nymphs and insects generally. They also eat crustaceans (Gammarus), small molluscs and plants. The structure of the mouth prevents it being drawn out into a downwardly opening tube as in the bream (which sucks up mud with its food) and silver bream seem to be much more selective feeders. They fast in winter, as do the bream.*

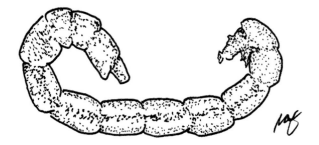

This commentary on the diet of the silver bream is echoed in the 1983 book *The Complete Freshwater Fishes of the British Isles* by Jonathan Newdick, which records:

> The food of the Silver bream is very similar to that of the
> Bream, Abramis brama and consists mainly of small bottom
> living animals supplemented by algae and other plant matter.
> It is unable to protrude its mouth when feeding as the Bream
> does and it barely feeds at all in the winter.

In his 1994 book *Freshwater Fish of the British Isles: A Guide for Anglers and Naturalists*, Nick Giles endorses the different feeding regimes of common and silver bream based on the structures of not only their mouths but also their gill rakers:

> Eddy Lammens, who works on the feeding ecology of bronze
> and silver bream in the Netherlands, has suggested that bronze
> bream feed most efficiently in fine silt/sandy substrates whilst
> silver bream, which have a coarser gill-raker system, are better
> at feeding in gravel-bedded areas. Age Braband, who studied the
> ecology of these two bream species in Lake Oyeren, southern
> Norway, found that bronze bream fed on invertebrates which
> were deeper down in the sediments than those taken by silver
> bream. Perhaps the two species split up the available habitat in
> these ways and thus avoid strong inter-specific food competition.

This observation about the structural differences in the gill arches and their effect on feeding behaviour is echoed by other research. C. Berg and colleagues published a study in 1994 (see Bibliography) based on X-ray cinematography, observing that common bream are able to adapt the gaps between their gill arches, creating a finer mesh size for filter-feeding. By contrast, silver bream are not able to reduce the channel between the arches. Common bream are thus preferentially able to feed on zooplankton.

As opportunists, silver bream also feed on available amorphous organic matter. This amorphous matter is known as detritus, apparently muddy-hued and unappealing but in reality a nutritious food source comprising a wealth of microbes inhabiting or digesting the decaying organic matter.

The statements by Alwyne Wheeler and Jonathan Newdick about silver bream fasting in winter may also be at odds with the experiences of some anglers. It is certainly my experience that silver bream can be caught in the winter, though often into dark as canal and pond waters clear in the cold. What is certainly true is that, like roach, rudd, common bream and similar coarse fish, appetite and feeding window may contract significantly in colder weather. This diet-suppressing effect may be exacerbated in the shallower, enriched waters that are favoured by silver bream.

THE SILVER BREAM LIFE CYCLE

Silver bream are shoal-spawners, adult fish of (generally) around 3+ years old gather in May to July around suitable vegetation, including submerged roots or sometimes on shallow gravel margins, typically when water temperature exceeds 15°C.

In his 1968 book *The Fishes of the British Isles and North West Europe*, Alwyne Wheeler notes:

> The silver bream spawns from June to July, often, though not always, intermittently at intervals of ten or eleven days, and two or three separate spawnings are made. Spawning takes place at temperatures of 16–17°C (60–2°F) amongst dense weed growth in depths of 2–3 ft (61–91 cm).

At this time of year, the heads and front half of the flanks of male fish become covered in rough white tubercles, thought to stimulate egg release by gravid females. The males jostle females near the spawning habitat, releasing their milt as the females release sticky eggs that adhere to the spawning surface. In his definitive 1968 book *The Fishes of the British Isles and North West Europe*, Alwyne Wheeler notes:

> Yellowish eggs measuring about 2mm in diameter are deposited by the female among dense plants to which they adhere.

In their 1967 *Collins Guide to the Freshwater Fishes of Britain and Europe*, Bent J. Muus and Preben Dahlstrom relate:

> Spawning is accompanied by boisterous displays and splashing, as with the bream. In southern part of its range it sheds 17,000–109,000 eggs in three spawnings at intervals of a few days.

The eggs released by any one female are typically fertilised by multiple males (polyandry) in group spawning, maintaining genetic diversity.

Silver bream exhibit no parental care once the eggs have been released and fertilised. Predation on the eggs by a wide range of fish and invertebrates is consequently very heavy. Alwyne Wheeler notes of the fertilised eggs:

> They hatch in five to eight days when the larvae measure nearly 5mm in length, and mature three or four years later then they measure from 10 to 13cm.

On hatching, larval silver bream are incompletely developed and immobile, adhering to submerged vegetation whilst they consume the yolk sac that remains attached to their undersides. Only when the yolk is fully consumed do larval silver bream become free-living, with subsequent growth rate strongly influenced by food availability and temperature. Hatching larvae are thought to be mainly nocturnally active.

Painstaking research conducted by my friend and colleague Dr Adrian Pinder into the early life stages of coarse fishes was published in 2001 as the still-definitive Freshwater Biological Association's *Keys to Larval and Juvenile Stages of Coarse Fishes from Fresh Waters in the British Isles*. Adrian observed and documented fish spawning and the subsequent development in captivity of collected eggs, emerging larvae and growing juveniles. Field observations corroborated that the eggs of silver bream are 2.0 mm in diameter, pale yellow in colour and attached to vegetation in shallow water at a minimum temperature of 18°C. His documentation, photography and scientific drawings of the subsequent development of hatching silver bream, along with virtually all other species of British freshwater fish, gave detailed insights into the characteristics of five larval and juvenile development phases, summarised here.

- Phase 1: free embryos, the first phase after hatching when the yolk sac is still attached, and the fins and gut are not yet developed. Length at hatching is 4.0 mm, the juveniles reaching a length of 6.0 mm by the time the yolk sac is fully absorbed. There are already distinctions with Phase 1 common bream, which have 23–24 pre-anal myomeres (muscle blocks along either side of the spine), whereas silver bream free embryos have 21–22.
- Phase 2: the larvae approaching 9.0 mm, the fins are not yet distinctly metamorphosed and the pigment on the back is not in distinct lines.
- Phase 3: intermediate larvae, in which the tail fin progressively becomes more distinct, with the pigment distinctly above the ventral aorta (a line beneath the head in front of where the gills join), unlike the common bream where this pigmented line is absent.
- Phase 4: older larvae up to 12.0 mm long with fins distinct, the pigmented line on the ventral aorta still distinguishing silver bream from common bream.

- Phase 5: young juveniles with fins progressively becoming more fully developed, the anal fin of the silver bream having no more than 24 rays when fully developed (whereas those of the common bream have more than 24 rays).

In his 1969 book *The Fishes of the British Isles and North West Europe*, Alwyne Wheeler notes of the growth of silver bream larvae:

The larvae at hatching measure about 4.8 mm. Growth in British waters is fast for the first two years and then falls off.

Alwyne Wheeler also notes:

Males mature in their third year, females a year later.

HYBRIDISATION WITH RELATED FISH SPECIES

Although Jonathan Newdick stated in his 1983 book *The Complete Freshwater Fishes of the British Isles* that

No hybrids involving the Silver bream have been recorded from British waters,

. . . this is not the experience of many other authors, scientists and anglers, least of all me!

In fact, silver bream hybridise readily and frequently with closely related fish species in many of the waters in which they occur. Many members of family Leucisidae with similar spawning habits tend to form hybrids. All are shoal-spawners, and hybridisation is particularly common amongst those that spawn communally on vegetation. These vegetation-spawning leuciscine fishes include roach, rudd, common bream and silver bream in Britain, as well as species such as vimba that occur in continental Europe.

The high incidence of hybridisation often associated with silver bream may be as much to do with the kinds of waters in which they thrive than it is due to greater promiscuity. As one example, silver bream often prosper in the sometimes murky still and nutrient-rich waters of canals that tend to have limited habitat diversity, with suitable vegetation for spawning often also occurring in restricted areas. There is consequently a tendency for shoal-spawning fishes of different species to converge on these limited vegetated areas as temperatures rise in the late spring, with inevitable cross-fertilisation.

I have certainly caught hybrids of silver bream with common bream, rudd and roach. These specimens can be tricky to identify, but the key external identification features tend to be intermediate between those of the parent fish, reproduced here.

Leuciscidae species	Spines/rays in anal fin	Spines/rays in dorsal fin	Scales along lateral line
Silver bream	III/21–23	III/8–9	44–48
Common bream	III/24–30 (usually 26–29)	III/9	51–60
Roach	III/9–11	III/9 11	42–45
Rudd	III/10–11	III/8–9	40–45

However, the identification of hybrids is further compounded by two other factors: mitochondrial DNA and back-crossing.

Mitochondria are the organelles within cells that produce energy through the breakdown and oxidation of food. Early in evolutionary history, mitochondria seem to have been separate unicellular organisms that became embedded symbiotically within larger cells. This explains why they have their own DNA (genetic material). Eggs cells contain mitochondria, but sperm does not. Consequently, hybrids carry more genetic material from the female parent than the male, not only affecting energy conversion but also influencing other features. Consequently, the features of hybrids, already tending to be quite variable, can vary further depending on which of the parents is male and which is female.

Back-crossing also introduces considerable complication. In essence, this occurs because hybrids are fertile and can cross back with other species or other fertile hybrids. A study by Billy Nzau Matondo and colleagues at the University of Liège published in 2012 (see Bibliography) compared 'sexual product quality' (egg size and semen density and consistency) in hybrids of common bream and silver bream. Whilst quality and fecundity were significantly lower than those of parental species, there was overlap demonstrating that hybrids have the biological capacity to produce high-quality gametes (reproductive cells) and thus a greater chance to produce F2 (second generation hybrid) and backcross (with true-strain parents) generations.

Matondo and team also studied hybrids of roach and silver bream in a study published in 2008 (again see the Bibliography). They found that eggs from female hybrids artificially fertilised with sperm of a corresponding hybrid male hatched and developed, female progeny producing eggs of high quality, although the semen of male hybrids was more dilute with lower sperm concentration than that of the parental species.

A study by Markus Vetemaa from the Estonian Marine Institute published in 2008 (see Bibliography) found that fertilisation and embryonic development success of hybrids between common bream and silver bream to the point of hatching was also high and comparable to that of pure parent species.

Other studies have found that survival rates of hybrids between roach, common bream and silver bream were similar to those of parent fish during hatchling and larval stages. Sexual activity by mature hybrids was also found to be high.

An interesting aspect of hybridisation is a comment by Izaak Walton in 'The Fourth Day: Observations of the bream; and directions to catch him' in his 1653 *The Compleat Angler*:

> Some say that breams and roaches will mix their eggs and melt together, and so there is in many places a bastard breed of breams, that never come to be either large or good, but very numerous.

Walton was not aware of silver bream as a separate species but may well have been referring to these morphologically distinct types of bream.

SILVER BREAM TAXONOMY

As noted earlier, the modern Latin name for the silver bream is *Blicca bjoerkna* (Linnaeus 1758).

In taxonomic terms, silver bream are part of the class of ray-finned fishes (Actinopterygii). Within this class, they are classified in the carp-like fishes (order Cypriniformes).

Within this broad order of carp-like fishes, the long-established family Cyprinidae (minnows or carps) covered some 3,160 species (in 376 genera) of varying forms occurring in fresh waters (only two species occur in fully marine waters) from North America, Africa and Eurasia. However, so diverse was this family of minnows and carps that it was ripe for revision based on contemporary morphometric and genetic evidence. Cyprinidae were split into many quite distinctly different subfamilies. One of these subfamilies is (or was) the Leuciscinae (the dace-like fishes).

Over recent years, there has been growing consensus amongst scientists that the long-established cyprinid family of fishes was too broad and that many of the subfamilies it encompassed were in reality a grouping of quite distinct families. A 2018 reclassification of the carps and minnows, based on morphological and genetic differences, split the large grouping into several distinct families. British examples of these newly redefined families include the revised Cyprinidae (true carps including common carp, barbel and crucian carp), Leuciscidae (minnows of Europe, Asia and North America including roach, common bream and Eurasian minnow), Tincidae (comprising the sole genus and species *Tinca tinca*, the tench) and Gobionidae (the gudgeons).

The family Leuciscidae, formerly the subfamily Leuciscinae, itself is diverse, comprising six subfamilies reflecting differing clades (groups of organisms believed to comprise evolutionary descendants of a common ancestor). These are outlined in the Table below.

Pogonichthyinae	North American clade
Phoxininae	True minnows
Leuciscinae	Old World clade
Plagopterinae	Creek Chub-Plagopterin
Laviniinae	Western clade
Pseudaspininae	Far East Asian (FEA) clade

The silver bream is part of the Leuciscinae (Old World clade) subfamily of the Leuciscidae (minnows of Europe, Asia, and North America). The Leuciscinae comprises 39 genera, spanning a wide range of species, including common bream, dace, chub, roach, rudd and the Eurasian minnow.

DISTINGUISHING SILVER AND COMMON BREAM

As observed in the earlier sections of this chapter, silver bream are often confused with small common bream. This is probably due mainly to a lack of familiarity with the distinctive features of silver bream, which are fairly obvious to people well acquainted with these fish. However, superficially, these two species of bream have many similarities, including their flattened profile, bright silvery flanks and long anal fin.

Exemplifying this similarity and potential for confusion, The Reverend W. Houghton wrote in his wonderfully illustrated 1879 book British Fresh-Water Fishes:

> It is hardly possible to distinguish this species from the young individual of the Common Bream . . .

Eric Marshall-Hardy added to this in his 1943 book Coarse Fish:

> There are anglers who doubt the existence of more than one species of Bream. I assure them, however, that, despite the very marked similarity between young and immature bronze and silver Bream, that the differences are actually material . . .
>
> It is only when Bream are young ('tin plates', 'bream flats') and small that the two species can be easily mistaken one for the other.

Alwyne Wheeler also notes in this 1969 book The Fishes of the British Isles and North West Europe that the silver bream

> . . . is frequently taken in the same waters as bronze bream (with which it is often confused).

Peter Tombleson also wrote in his 1954 book Bream: How to Catch Them:

> The young of the Common Bream, with which we are mainly concerned, are often taken for the adult Silver Bream but one sure method of ascertaining the species is to compare the anal fin. That of the Silver Bream is shorter. Its scales too are larger.

To this, A. Lawrence Wells wrote in his 1941 book *The Observer's Book of Freshwater Fishes of the British Isles*:

> *Although the White or Silver Bream is a much smaller fish than the Common Bream – rarely reaching a weight of a pound and a half – the young of the latter fish are often taken for adults of this species and vice versa.*

Many other authors besides have commented on the potential confusion between silver bream and younger common bream. Nonetheless, once you get your eye in, the distinctions are quite clear. These have largely been addressed in detail in preceding sections of this book but are summarised in graphic form here to aid the observer. The first pair of images examines the heads of the two bream species as the large eye and non-protruding mouth of the silver bream are amongst the more obvious distinguishing features. The second pair of images focuses on distinctive features of the body and fins.

Based on genetic analysis assessing differences between European cyprinid species, a study by Häunfling and Brandl published in 2000 suggested that the genus *Blicca* (the silver bream is the only member of the genus) should be merged with the genus *Abramis* (containing only the common bream). This proposal did not take root, though was reflected in a limited number of publications around the time. However, it does emphasise the genetic and also morphological similarity between the two species.

One of the diagnostic means for distinguishing cyprinid fish species is to dissect out the pharyngeal bones (the 'pharyngeal teeth') found in the throat, modified from gill arches. Silver bream have pharyngeal teeth in two rows with 5 + 2 on each side, whereas those of the common bream are in a single series with five on each side. However, identification by pharyngeal teeth is a job for experts as it entails killing and dissecting the fish.

Head of silver bream

Eye diameter equal to or
greater than snout length

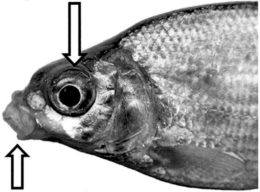

Lips protrude to a
small extent, similar to
the mouth of a roach

Head of common bream

Eye diameter smaller
than snout length

Lips protrude prominently,
forming a cylinder for
probing into soft sediment

Body and fins of silver bream

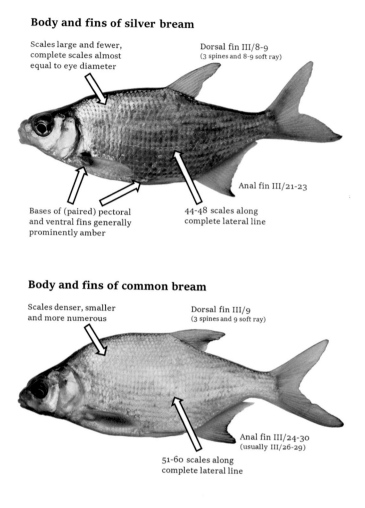

Scales large and fewer, complete scales almost equal to eye diameter

Dorsal fin III/8-9
(3 spines and 8-9 soft ray)

Anal fin III/21-23

Bases of (paired) pectoral and ventral fins generally prominently amber

44-48 scales along complete lateral line

Body and fins of common bream

Scales denser, smaller and more numerous

Dorsal fin III/9
(3 spines and 9 soft ray)

Anal fin III/24-30
(usually III/26-29)

51-60 scales along complete lateral line

PREDATORS OF SILVER BREAM

I am sure that many anglers will have shared my experience of a pike taking a struggling silver bream as it is retrieved when hooked. Silver bream are in fact a nice bite-sized snack for a large-mouthed predatory fish like a pike. Zander too will readily take silver bream.

In fact, smaller silver bream, as indeed most other small coarse fish species or their juveniles, are acceptable prey for a very wide range of predatory animals.

These predators include a diversity of piscivorous and opportunistic fish species such as perch, chub and eels. Many piscivorous birds, such as kingfishers, herons, egrets and cormorants, will also opportunistically take silver bream, and so too may predatory mammals such as otters and mink.

Okan Yazıcıoğlu and colleagues in a scientific 2017 paper (see Bibliography) note of the silver bream:

> . . . it is an important food supply for predator species inhabiting aquatic habitats.

MORE INFORMATION ABOUT SILVER BREAM

If you want to know more details about the biology of silver bream and other British freshwater fishes, I recommend three of my other books, both also listed in the Bibliography:

- *The Complex Lives of British Freshwater Fishes.* (2020). CRC/ Taylor and Francis, Boca Raton and London.
- Everard, M. (2013). *Britain's Freshwater Fishes.* Princeton University Press/WildGUIDES, Woodstock, Oxfordshire.

Three

In his 1969 book *The Fishes of the British Isles and North West Europe*, Alwyne Wheeler wrote:

> *Silver bream have little value for anglers, their small size being a disadvantage. They are not fished for or eaten for the same reason. Their competition for food with the bream in its immature years may make their presence undesirable if large bream are being deliberately cultivated.*

In their 1967 book *Collins Guide to the Freshwater Fishes of Britain and Europe*, Bent J. Muus and Preben also wrote:

> *Its small size, and the number of bones, decreases its value both for food and as a sport-fish.*

I guess that all, or at least most, readers of this book will disagree! All species have their virtues and fans, and all, including the silver bream, can inspire and have inspired angling interest.

As with most of my books on fish and fishing, reflecting the way I approach catching specimen fish of any species by design, I follow the advice of the great pioneering angler Richard Stuart Walker. Walker is seen as the 'father' of modern specimen angling as, until the 1950s and 1960s, the general assumption amongst the angling community was that the capture of a specimen fish was a matter of chance.

DOI: 10.1201/9781003311041-3

Walker applied a scientific approach, innovating techniques, baits, timing and all other aspects of his approach to the capture of big fish by design. This resulted in a remarkable tally of record and specimen coarse and game fishes. His approach was also articulated with clarity in his engaging and prolific writings. One of the key foundations of Walker's innovative approach was a set of three steps: location, bait and presentation.

The sequence really matters. If the fish you hope to catch are not present, you will not catch them. Nor will you if they are not feeding, or can be induced to feed. All other approaches are unlikely to bear fruit if location is not first addressed. Location is a matter of being on the right water, at the right time and at the right depth and in the right habitat element. Then, it is a matter of presenting a bait upon which the fish will feed. Only then does presentation come into play, to put the right bait in the right place with subtlety but also sufficient capability to detect bites and play the fish when hooked.

SILVER BREAM LOCATION

Like many species in the family Leuciscidae, silver bream are opportunistic feeders adapted to a wide variety of still or slow freshwater habitats, as well as some brackish lower reaches of rivers. Silver bream also do well in many canals as well as lakes and ponds. In rivers, silver bream are to be found in slacker reaches, including out of main flow in the margins of larger rivers such as the Thames.

Generally, silver bream feed at or near the bed of the river, canal or pool. Consequently, this is where to fish for them; a bait presented on the surface or up in the water column is far less likely to attract their attention. However, bear in mind also that silver bream are known to come into shallower margins as darkness approaches, so in these instances presenting a bait on or near the bottom may mean fishing apparently shallow.

Amongst the temporal aspects of silver bream location is notably crepuscular behaviour in clearer water conditions, though these fish appear to feed freely throughout the day in murkier water. Some authors have stated that silver bream cease to feed in the winter, though this is not automatically the case. An eye on longer-term weather trends is also a useful location aid, as a rapid decline in water temperature can switch off many fish species including silver bream. In rising temperatures, silver bream and many other fish species are likely to come onto the feed regardless of the absolute temperature. Likewise, fish acclimatise to stable temperatures and can be expected to feed even in cool water, though the feeding window may be narrow.

SILVER BREAM BAITS

As omnivores, silver bream have truly catholic tastes. Consequently, many baits can be deployed to catch silver bream. As opportunists, a little loose feeding can help not only bring them onto the feed but also seek out the bait one is presenting on the hook.

Reflecting this broad palette, and addressing both common and silver bream together, Arthur P. Bell wrote in his 1926 book *Fresh-water Fishing for the Beginner*:

> *The best baits for bream fishing are worms (preferably lobs), paste, breadcrust, gentles, or wheat, the first named being generally the most useful.*

Insect larvae comprise an important natural element of the diet of the silver bream. Consequently, insect larvae are a bait immediately recognisable to and accepted by silver bream. These include maggots in all of their diversity of types, including maggots dyed in various colours, gozzers (large maggots), and the smaller pinkies and squats. So too can their pupae, known as casters. For some species, match anglers favour dead maggots over live ones, though whether this is true for silver bream is not certain.

The larvae of chironomid midges, red and worm-like and living in profusion in muddy river and lake beds, are known as bloodworms. Bloodworms are known by match anglers to be a highly effective bait for many species of freshwater fish. However, as procuring bloodworms is laborious or expensive, and presenting them on the hook is specialised, they are banned in many matches, and their use is beyond the guidance of this little book.

Another insect-based bait that was common when I was a child, but less commonly available now, are wasp grubs. These grubs are extracted from wasp nests from which the adult wasps have been driven away, generally by smoking them out. The grubs are extracted in the papery cells in which they grow in the nest. Hooked individually, they are a tasty bait attractive to many fish species.

As Arthur P. Bell informed us, worms – true worms and not insect larvae like bloodworms – are a fine and effective bait for silver bream and many other freshwater fish species. Lobworms, Britain's largest earthworm species, are very effective baits for many freshwater fish species. However, they may need to be cut or broken into smaller sections to enable smaller species such as silver bream. Smaller worms can be as or more effective. These include red worms, known as dendrobaena, commonly found in leaf litter. The rather smellier brandlings, striped redworms that exude a yellowish liquid, can also be used as an effective bait. Although I had formerly assumed that the smelly brandlings were less palatable to fish, I discovered that I was wrong during a campaign in 2021, finding them highly effective for catching many silver bream, perch, ruffe, roach, rudd, common bream, gudgeon and eels. Some anglers that I know recommend worm/maggot cocktails for silver bream.

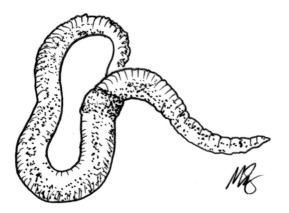

Other animal-based baits favoured by silver bream, amongst other fish, include shrimps or prawns available from supermarkets. Freshwater shrimps too will be taken but are hard to collect and use in practice.

As opportunists, silver bream also eat molluscs, and even occasionally fish eggs. There is increasing use of fishmeal-based pellets by anglers; silver bream and many other fish freshwater species will readily identify and ingest these novel baits, particularly where they are introduced in quantity by carp or match anglers seeking larger fish.

Whilst on the subject of animal-based baits, I will share with you an absolutely killer silver bream bait that I discovered quite by accident. The bait in question is luncheon meat, with 'bacon grill' seemingly the most attractive. Blocks of luncheon meat, opened fresh from the can or stored in the fridge (for short periods) or freezer (for longer periods), are diced into cubes. Finely diced or grated luncheon meat can be used as a loose feed to attract fish. Slightly larger cubes, sized as appropriate to hooks and anticipated quarry, are diced or ripped from bigger chunks of luncheon meat. My discovery about the attractiveness of bacon grill to silver bream came about by chance after returning from a barbel fishing trip with excess bait and setting up a lift method rig on an 8 lb line and a size 8 hook cast close to my far finer silver bream rig anticipating intercepting a roving tench. However, the luncheon meat rig flew away almost immediately, yielding me a specimen 1 lb silver bream. There followed a stream of about 20 more voracious silver bream, some of which hit the bait 'on the drop' even on heavy and fast-sinking lift method tactics!

The willingness of silver bream to ingest amorphous organic matter known as detritus means that other soft, vegetable-based baits are accepted readily.

Bread is a superb bait for silver bream and many coarse fish species. My own strong hook bait preference is bread flake, which wafts enticingly with neutral buoyancy if mounted on a light, fine-wire hook. In the summer, and in clear water conditions, punched bread is a great alternative, particularly when presented on the pole. Bread paste is also a staple bait, though not as commonly used today as formerly, and it has the added bonus of readily taking flavourings or other additives (dyes, cheese, etc.) that can be mixed in as dampened bread is kneaded into a paste.

Other, more exotic types of pastes, such as those marketed for carp, barbel and other types of fishing, can also find a place in the silver bream angler armoury, though arguably they are unnecessary except perhaps on waters where these smaller fish become preoccupied with larger baits fed to attract larger fish species. The same is true of fragments of boilies.

Sweetcorn is another bait readily accepted by silver bream, as indeed many coarse fish species. My preferred method is to loose feed with liquidised corn containing a few partly cut pieces of corn, the fish attracted by the scent of the bait and selectively picking up whole grains on the hook. Although I have had a lot of success with presenting buoyant plastic corn on the hook for tench, I have not found it so effective when targeting silver bream. Stewed wheat is another bait that is less commonly used today but which is attractive to silver bream and many other coarse fish species, amongst a range of other vegetable-based baits such as sections of pasta.

When fishing for silver bream, overfeeding with ground-bait can be problematic as it tends to attract larger fish species such as common bream and common carp. However, fed sparsely, a little groundbait can attract the attention of silver bream. Small quantities of groundbait are best introduced in a targeted way, for example via a pole cup, bait dropper or swimfeeder, or else in an accurately thrown dense ball. It is important that the bait is dense enough to sink rapidly, only breaking up on the bed of the river, pool or canal so as not to be intercepted and attract the attention of other fish species as it falls through the water column. For active particle baits, such as maggots or small sections of worm, a swimfeeder will release these lively morsels in the direct vicinity of the hook bait.

Loose feed or groundbait need not be complicated so long as it is matched to the hookbait. When maggot fishing, use maggot groundbait, which also works well with worms or worm/maggot cocktails on the hook. Likewise, liquidised bread is highly effective for offerings of bread on the hook; I often add a handful of layers mash (liquidised chicken feed available from pet food outlets) to dampened liquidised bread to give the mix a bit more flavour as well as additional bulk and cohesion for throwing, also increasing overall density, enabling balls of groundbait to sink rapidly to the bed before breaking up.

Little and often is best to keep aroma in the feeding zone, but without either attracting an influx of bigger, competitive fish or overfeeding the silver bream.

PRESENTATION WHEN SILVER BREAM FISHING

Once the all-important element of location has been attended to, and appropriate baits selected to attract the attention of feeding fish, we then get to the third element: presentation. Reaching for the favourite or habitual kit without having taken full account of the first two steps of location and bait will otherwise hamper you right from the outset

It is assumed at this point that all readers will have a general knowledge of angling techniques. The focus then is on how to deploy and perhaps tweak them for the task of extracting silver bream in different situations.

Remember that silver bream are fish that feed predominantly on the bed, or close to it. This influences all aspects of angling for these fishes.

Also remember that even specimen silver bream are far from the most massive of fishes, or the fussiest of feeders. It would be overstating the issue massively to suggest that there are specialist silver bream rigs, stepped-up rods and other dedicated kit to focus your attention or trouble your wallet. But good basic technique and care with presentation are assumed. I have, in fact, found no writers claiming anything more specific.

Pole fishing is one of the finest and most precise methods for offering a bait to potentially feeding fish in still and slower-flowing waters where they can be expected to be feeding, or be induced to feed, within range of the pole. No specialist pole tackle is required, though the plumbing of depth is particularly important. Match anglers, of course, know this well already, but this detail is often less well appreciated by casual pleasure anglers and specimen fishermen. It is advisable to use a wide-bottomed 'silt plummet' that will settle on soft sediment rather than sinking into it, giving a potentially incorrect indication of where the bait will be presented. It is also important for this same reason to allow the plummet to settle gently onto the bed. The depth should be set such that the hook is an inch or so above the sediment, amongst the angler's free offerings but wafting enticingly and availably above them so that the fish will be most likely to encounter and intercept them. Remember that the water level in canals can change as locks are opened, so you may need to make periodic adjustments to stay in touch. I favour a No.7 elastic through the top sections of the pole, elasticity to absorb the struggles of the fish with sufficient 'backbone' to control the fish and avert the elastic stretching completely beyond which point it has no further give and can result in the fish breaking the line or more likely pulling free of the hook. I also favour a hook link of around 1 lb 8 oz (around 0.7 kg) below the bulk weight. A well-balanced pole setup absorbs considerable strain, averting the need for heavy lines that also hamper optimal presentation. Also, in the event of needing to pull for a break if snagged, a pole float (particularly one with a metal stem) can turn into a dangerous weapon when pinging back on a fully stretched elastic. For this reason, always aim to pull with the float submerged to buffer the release of energy and also turn your face away from a snagged rig. It is customary to fish an olivette as a bulk weight at the bottom of the main line but above the lighter hook link, rapidly taking the bait down to the target depth. However, my preference is to use a swivel instead as this not only has weight but also aids in joining the main line to the lighter hook link and can address tangling if fish spin when played.

Another subtle and appropriate method for presenting a bait to silver bream, and one of my favourites too, is the lift method. This method was formerly far more commonly used, particularly for finicky tench. It entails attaching a waggler float to the reel line at a suitable depth but with a heavy weight where the reel line joins a short hook link of just a few inches. Although the weight may appear unsubtly heavy, if critically balanced against the buoyancy of the float, it effectively becomes neutrally buoyant and will not be felt by a fish picking up the bait. The weight is generally recommended to be just sufficient to sink the float, such that when the fish picks up the bait the float lifts, even falling over flat. In fact, in about half the cases, the float may sink as the fish moves off with the bait in its mouth. Once again, my preference differs a little in that a heavy swivel can be substituted for a weight, also assisting in joining the reel line to the hook link. I also prefer to use slightly less weight such that the float is just buoyant enough to lift it and a dot down the float by winding back slowly against the friction of the swivel on the river, canal or pool bed. This use of friction rather than sheer weight enhances further the sensitivity of the method when a fish disturbs the bait and also enables the reel line between the float and swivel to run at a more acute angle, presenting less of an obstruction to fish moving near or over the bait. My preference is also to use a centrepin reel for optimal control and enjoyment on the play, but that is purely personal as any open- or closed-face fixed spool reel will suffice. I also like to set my waggler up as a sliding float enabling me to explore different depths.

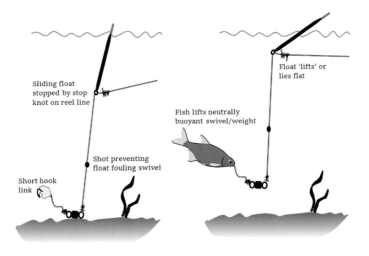

Silver Bream Fishing

There is no need to take these relatively more sophisticated approaches for, let's face it, a less-than-sophisticated fish that is hardly hugely pressured by dedicated specialist anglers. A standard waggler float approach can also be effective in still and slower-flowing waters. In running waters, stick or Avon float presentation can be effective depending on the strength of flow, depth and distance. On a stronger-flowing river such as the Thames, I catch silver bream on the heavier Avon setup as it offers the best presentation at depth. In slacker flows, the stick float or indeed a heavier waggler may offer the best and most sensitive presentation. In all cases, working the float slowly through the swim by holding it back is important, searching out the nooks and crannies of the bed for where a hungry silver bream or other fish may be lurking; this advice is of particular importance in cool water when the fish may be far less active.

Leger tactics can also be effective in all waters, particularly into dusk or dark when floats become less visible. Luminous floats are available, but my own experience of them is that bite detection with appropriate leger setups is far more sensitive and also strains the eyes to a far lesser extent.

In terms of terminal leger tackle, my strong preference is to use small swimfeeders placing small amounts of bait close to the hook bait on the principle of feeding 'little and often' on each cast. It is vital to ensure that there is minimal friction in the terminal end of the rig such that a fish picking up the bait feels little or no resistance. My own preference here when fishing for silver bream, roach, dace and chub is to tie the feeder onto a loop of 4 lb nylon with a captive swivel, which effectively acts like a zero-friction paternoster. The captive swivel is tied into a loop of reel line within which it slides freely. At the terminal end of the loop of reel line, three double overhand water knots are tied to create a slightly stiffer end from which the attached hook link is attached loop-to-loop, standing out slightly and thereby reducing the risk of tangles on the cast. If any one element of this 'multiple redundancy' rig tangles, other elements of the paternoster/sliding leger still operate meaning that each and every cast can be fished effectively even if the tangle is only noticed when the rig is retrieved. Although the swimfeeder or bomb thus presented runs in a loop of reel line, there is no risk of tethering fish on a break-off as the lighter hook link breaks if snagged and the loop of line breaks if the feeder is snared.

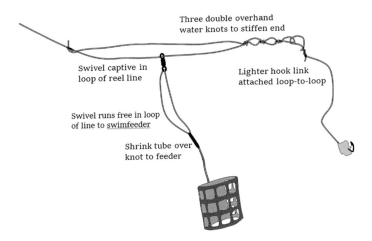

Three double overhand water knots to stiffen end

Swivel captive in loop of reel line

Lighter hook link attached loop-to-loop

Swivel runs free in loop of line to swimfeeder

Shrink tube over knot to feeder

Bite detection when legering can be achieved using a number of different methods. In order of increasing sensitivity, I would choose a bobbin, swingtip, touch legering or quivertipping. Attaching a bobbin to the line between rod rings can be effective when expecting a low frequency of bites and relying on electronic bite indicators, though there is a trade-off with sensitivity, albeit that this does not matter if using self-hooking rigs. Swingtips were formerly far more commonly used, though, in my experience, they offer no advantages over a sensitive quivertip. Touch legering is very sensitive, the rod pointed at the bait averting friction of line on the rod rings, and your fingertips on the line offer all sorts of information about what is happening to the bait, though this does require constant attention and your fingers may lose sensitivity in the cold. Consequently, my favoured bite detection method is generally a sensitive quivertip of the lightest test curve available (ideally ½ or ¾ ounce). I modify my quivertips by removing the tip ring with gentle heat from a lighter, feeding a silicone float adapter onto the tip before regluing the tip ring, then inserting a betalight into the float adapter. The adapter/betalight works well during the day by presenting a bigger target to watch, and obviously into darkness due to its visible glow. I always pair up my quivertips. This is but anticipating doubling my chances (Sod's Law says that bites always come on one rod when you are busy baiting the other!), but as the two tips act as a reference for each other, also better distinguishing the actions of fish from wind and current. In my early days experimenting with illuminated rod tips, I often missed 'massive bites' into darkness when fishing with just one rod; only far later did I realise that the 'bite' was in fact an artefact of rapid eye movement as I was nearly nodding off in the darkness!

The bulk of silver bream fishing need not be at distance nor require self-hooking rigs. However, when either or both of these factors are required, the helicopter rig is highly recommended. As a solid endorsement, it is recommended by a former holder of the British Rod-caught Record silver bream. Helicopter rigs may look complicated, particularly to anglers like me who predominantly use traditional techniques, but they need not be. One of the principal benefits of the helicopter rig is that it averts tangles on the cast as the hook link is free to rotate, like a helicopter's blades, around the main line connecting the rod/reel to the terminal weight of swimfeeder. The short hook link connects to a swivel that is free to rotate, held between two beads on the main line, generally with some anti-tangle tube preventing the hook link fouling the line below the lower bead. The distance from lead or feeder to the hook link can be adjusted as appropriate to conditions: shorter to maximise self-hooking effect but potentially extended to prevent the baited hook from being drawn into soft silt or weed as the weight sinks. Ideally, the baited hook should be light such that the hook link straightens out as the weight or feeder sinks, further minimising the risk of tangles and also enabling the bait to waft enticingly when disturbed by the movements of fish. When the desired fish are at range, the helicopter rig with a swimfeeder of sufficient weight to self-hook the fish can be effective for presenting baits such as maggots or worm/maggot cocktails.

I have not mentioned much in the way of line strength and hook size in the aforementioned descriptions. Whilst these are not critical, these are a matter of choice according to the experience of the angler, the terminal gear they choose, the baits they use and the circumstances in which they are fishing. Subtlety is important for presenting to this smallish fish, but tackle also needs to be sufficiently strong to play out any fish you encounter, be that silver bream or other species likely to pick up the bait in the places you are fishing. I have in fact enjoyed some memorable captures of sizeable tench, eels, roach and common bream when silver bream fishing, and they are all very welcome additions to the session!

LURE FISHING FOR SILVER BREAM

Lure fishing has evolved massively in British angling over recent years. Formerly, lure fishing may have been considered as something limited to larger predatory freshwater fishes such as pike, perch and zander as well as trout and salmon and, in warmer weather, opportunistic omnivores such as chub.

A new generation of micro-lures is changing former assumptions. The reality is that virtually all freshwater fishes are carnivorous to varying degrees, particularly when their metabolism is raised by elevated temperatures. Most graze on invertebrates of varying sizes and are far from averse from opportunistically predating on fish eggs and fry. Generalist imitative fly patterns work well for dace throughout most of the year and for roach, rudd and other coarse fish species in warmer months. Silver bream and common bream feed extensively on bloodworms, the blood-red larvae of chironomid midges abounding in rich sediments, as well as other available invertebrates. Consequently, it should not really be surprising that they can be enticed to take small lures imitating or approximating these types of food items. These micro-lures are most effectively fished near features such as canal, bridge and lock walls, wooden debris and submerged vegetation. It is a fun way of fishing!

I have caught well on micro-lures, but mainly perch, ruffe and chub. I have yet to catch a gudgeon – a serious personal target! – or a silver bream on a micro-lure. However, my angling friend, Richard Widdowson, is a dedicated and expert proponent of the use of tiny lures of an inch or less in length, presented on small hooks with tungsten bead jig-heads. He targets and catches gudgeon as well as a huge diversity of other freshwater and sea fishes not normally associated with lure fishing. From tench, roach and carp to eels, ruffe and both common and silver bream, Richard has caught freshwater fish of many species using these tiny lures, either on small jig heads or by drop-shotting.

Richard has had great success with micro-lures with many freshwater fish species, though he admits that he has caught only a handful of silver bream this way, noting:

> They did take a small green paddle tail lure so I'm guessing they are drawn to that colour, on a small jig head.

Lure fishing for silver bream may not then be the most productive approach, but it is one that perhaps deserves more attention. It is one I intend to try more diligently for silver bream, along with my pursuit of gudgeon and other of our more neglected freshwater and marine fish species.

THE FIGHTING SILVER BREAM!

Common bream are renowned for their lack of fight when hooked. This is perhaps unfair regarding common bream inhabiting faster rivers, which not only are less slimy than their still water counterparts but also know how to 'kite' in the current using their broad bodies.

Silver bream tend to get tarred with the same 'non-fighter' brush. For example, Eric Marshall-Hardy wrote in his 1943 book *Coarse Fish*:

> Bronze Bream are poor fighters and 'Silvers' are even worse, but, if it is possible, they make up in quantity what they lack in sporting quality when 'on the feed'.

However, this is not fair, perhaps due to the low discrimination amongst anglers between the two freshwater bream species.

It is my experience that I can tell a silver bream when I have hooked one due to its fight. They use their broad bodies to 'kite' in the current-like common bream and can also 'thump' against the rod tip when kicking side-on as do roach and common bream. However, with silver bream there is also the sensation of 'fluttering' in the fight as the fish ripple their body actively, much as a dace, though often as a right angle to the rod as they also use their body surface area to resist the pressure. Whilst not putting their fight into the same category as a mahseer or barbel, silver bream nonetheless do put up a respectable and characteristic account of themselves.

PHOTOGRAPHING SILVER BREAM

Taking a good photograph of your specimen silver bream can be tricky, especially as fish care has absolutely to be a priority. If you can't photograph the fish without putting it under undue stress, get it witnessed if need be, then simply pop it back into the water and store away the memory instead!

Grayling are tricky to photograph as their muscular bodies seemingly never stop writhing. Dace are tricky too as they are round-bodied and hard to hold side-on, but also very bright, meaning that the image is prone to bleaching. Though flat-bodied, silver bream are, at least from the point of photographic exposure, troublingly silver! In my experience, they are by far the most difficult freshwater fish to photograph clearly as their brightness tends to bleach out detail in the photograph.

This is not a book dedicated to photographic skills. My assumption is also that most anglers seeking guidance will be using mobile phones or compact cameras rather than more specialist photographic equipment.

The first advice then is to turn the flash off, as otherwise you will be seen holding a white disk due to the intense reflection! Ideally, set the camera up, or get a colleague to hold the camera, looking slightly down on the fish as that way you avoid direct reflection from the brilliant flank. Also, if you have that capability, set the exposure onto the fish, not the background. Or, if you can't control exposure, seek out a lighter background. Digital cameras are great from this point of view as you can instantly assess the clarity of the image. If you have to use flash, photographing from above the fish becomes even more important.

If you need to take detailed photographs for identification, take a large number of images of the fish, again orienting slightly downwards and with the focal point on the brighter part of the flank to let the camera adjust exposure automatically to avoid bleaching the image. Remember to take plenty of close-up photographs of key identification features, such as of the mouth and eye with the protrusible jaws at maximum extension, and also the lateral line. Also, take close-up photographs of the extended dorsal and anal fins, ideally with the light coming through them or at least on a white background to enable the detailed counting of fin spines and rays. The photograph I took of my personal best silver bream at the time of writing, a 1 lb 4½ oz lying on a wet landing net head, came out well when exposures were carefully controlled simply by shifting the focal point on my mobile phone.

More detail with camera angled slightly downwards, avoiding glare from fish

Angle all wrong, reflections from silver flanks bleaching out details

Silver Bream Fishing

A reminder again: fish care must come first. If you need to extend the photographic session, allow the fish to recover periodically in a soft net or pike tube. Then release the fish as promptly as possible, checking first that it is fully recovered and able to maintain correct orientation so that it can swim off strongly, evading any predators that may be lurking nearby.

SILVER BREAM RECORDS IN MODERN TIMES

The British Rod-caught Record silver bream at the time of writing – possibly one that may be beaten by the time you read this book! – is a fish of 3 lb 4 oz (1.47 kg) caught on 21 May 2012 by Gareth Evans from Mill Farm Fishery, Bury, near Pulborough in the county of West Sussex. Gareth Evans caught the fish on a small pellet offered on scaled-down carp tactics, reasoning that the biggest silver bream would have adapted their diet to exploit the abundance of highly nutritious baits introduced by carp anglers.

To set in context how monumental this current record is, we have to bear in mind that the British Rod-caught Record silver bream set in 1998 was a fish of 'only' 15 oz. This fish was taken from Grime Spring, Lakenheath in Suffolk in eastern England, by Dennis Flack. Dennis Flack had made something of a name for himself by targeting smaller species, holding several records.

Dennis Flack's record fish stood for five years. Amazingly, all subsequent records have been held by fish taken from just one fishery: Mill Farm Fishery in Bury, near Pulborough in the county of West Sussex in south-east England.

This sequence of silver bream records from Mill Farm Fishery were:

- 1 lb 4 oz, spring 2003
- 1 lb 11 oz, spring 2003
- 1 lb 12 oz, captured by Michael Davidson on 26 April 2003
- 1 lb 13 oz, caught by Duncan Charman on 14 April 2005
- 2 lb 2 oz, caught by Christine Smith in 2005
- 2 lb 8 oz, banked by James Waters in December 2008
- 2 lb 11 oz, by Doug Plant in 2009
- 2lb 14 oz, landed by Phil Morton on 8 May 2009
- 2 lb 15 oz, caught by Gary Barnett on 9 October 2011
- 3 lb 1 oz, landed by Matthew Faulkner on 12 May 2012

I know a little of the details of Duncan Charman's former 1 lb 13 oz record because we have talked about it! Duncan took this impressive fish from Mill Pond, one of three lakes on the Mill Farm Fishery. It was captured by pole fishing at a distance of 9–10 metres with worms and maggots. Small amounts of free offerings were introduced by bait droppers to stop carp from intercepting them as they sank. Duncan had found that the silver bream were mainly to be found in open water, neither in the deepest nor in the shallowest water but at intermediate depth. The encounter with the 1 lb 13 oz silver bream on that fateful day – 14 April 2005 – put Duncan into the annals of angling history. Duncan records what he wrote in his angling diary in his 2012 book *Evolution of an Angler*:

> At 11am, unbeknownst to me, a big silver had entered my
> swim and was soon to find my worm and maggot cocktail.
> As soon as she was hooked, I knew it was something special

and after a brief fight she dropped over the drawstring before
being weighed. At 1lb 13oz 8drams she recorded a new British
record, something deep down I was aware was about to
happen. The day wasn't over and an hour later I caught another
big silver bream, this time weighing 1lb 6oz, a brace I won't
forget in a hurry.

The ratification process took a long time; the fish rounded down to 1 lb 13 oz with the 8 drams discounted. The final issue of the certificate occurred only after another larger fish had been caught from Mill Farm and was being put through the British Record claim process.

Duncan Charman has subsequently had several bigger Mill Farm fish, including on 26 April 2013 the biggest known British brace of silver bream weighing in at a hefty 3 lb 0 oz and 3 lb 4 oz. Duncan has kindly allowed his photograph to be included in this book.

WHAT'S SO SPECIAL ABOUT MILL FARM?

With such an impressive sequence of British Rod-caught Record silver bream coming from this single West Sussex fishery, one has to ask what is special about the Mill Farm

Fishery that not only suits silver bream but allows them to grow to prodigious and unprecedented sizes.

Ostensibly, there is nothing extraordinary about the three lakes now to be found at Mill Farm. All are man-made, dug into former grassland as part of development by owner Jeremy Stuart Smith, a beef farmer and supplier of plants and trees, on part of the 120-acre Bury Mill Farm. Central to the design was the creation of a wooded lake which would attract wildlife. Run-off from adjacent farmland feeds into the lake, but the fishery is supplied with water from its own borehole maintaining oxygen levels.

The Mill Pond pleasure fishing lake and the Specimen Carp lake are the oldest of the three pits now on site, realising the landowner's vision of angling and nature conservation working together. The third lake, the Hammer Pond, opened for pleasure fishing some years later in 2000, in part to reduce pressure on the other two waters. Mill Farm imposes a close season from New Year's Day until Good Friday to further reduce pressure on the wider fishery ecosystem, including its fish populations. There are also restrictions on the use of groundbait, boilies or method feeders as well as keepnets. Furthering the conservation ethos, night fishing is not allowed between sunset and 7:00 am.

All three lakes have produced notable fish featuring in the angling press. Mill Farm Fishery's greatest claim to fame, though, is undoubtedly the sequence of British Rod-caught Record silver bream, attracting anglers as there are believed to be bigger fish still to be caught. Mill Farm is without doubt an interesting fishery. However, it is hard to pick out any distinctive features that explain why all three lakes have produced large silver bream. When I fished the lakes at Mill Farm on invitation of the fishery manager in April 2022, I caught many silver bream and other fish but could not discern anything of particular significance about the water bodies themselves to explain their unique place in angling history beyond their shallow profiles and high productivity. Duncan

Charman, former silver bream British Rod-caught Record holder, is very familiar with the venue and believes that the absence of competing common bream is one of the most significant features. This is an entirely plausible idea as common bream not only compete strongly for food with silver bream, potentially suppressing their maximum growth, but also tend commonly to hybridise with them diluting the purity of the genetic strain. Roach, some up to 2 lb, are present in Mill Farm too, though they do not seem to be as competitive or to hybridise regularly.

SILVER BREAM RECORDS IN FORMER TIMES

The list of British Rod-caught Record silver bream above dates back to Dennis Flack's 15 oz fish in 1998. However, larger fish were formerly registered as records.

Angling records within the United Kingdom, Northern Ireland and the Channel Islands (collectively the 'British Records') are today officially the responsibility of the British Record (Rod Caught) Fish Committee (BRFC). The BRFC was formed in 1968 and since 2009 is part of the Angling Trust. One of the first and most controversial decisions of the BRFC was to purge the pre-existing British Record list of former records that could be verified with photographic evidence, witnesses, tested weighing scales, correct species identification and other supporting evidence. One of the many casualties of this purge was a former record silver bream of 4 lb 8 oz. Some other former 'record' and other recorded notable silver bream are also substantially in excess of the modern British Record.

Alwyne Wheeler documents in his 1969 book *The Fishes of the British Isles and North West Europe* documents:

The record British rod-caught fish weighed 4½ lb (2.04 kg).

In all probability, Wheeler refers the same fish documented by Eric Marshall-Hardy, then-editor of *Angling* magazine, in his delightful little 1943 book *Coarse Fish* written during the Second World War:

> The record Silver Bream (Blicca bjoerkna) was taken at Tortworth Lake (Glos.) by Mr. C. Rhind in 1923 and weighed 4½ lb.

Eric Marshall-Hardy also records:

> Other fine specimen Silver Bream:
>   Feb., 1922. R. Thames (Egham). Mr. G. Burwash . . . 4 lb.
>   July, 1933. R. Derwent (Yorks.). Mr. J. Bowater Two of 4 lb.

For those who 'have their eye in', silver bream are distinctive. However, their similarities with smaller common bream and the frequency of hybridisation can reasonably be assumed to have given some scope for misidentification. Despite this, there is no reason to assume that anglers of bygone eras were any less observant of the details of the fish they caught. Nonetheless, these former record fish are very substantially in excess of the normal run of the mill of British silver bream encountered today, and even of the sequence of recent British Rod-caught Record fish from exceptional waters. Perhaps, these formerly recorded fish were 'real' silver bream, thriving in rich waters and in the absence of competition from other now widely stocked species such as common bream and common carp. What is certain, though, is that the burden of proof in former times was far lesser than that required today.

Also certain is that the specimen credentials of the silver bream are still underplayed in Britain with respect to those of other freshwater fish species. At the time of writing (the start of 2022), the Angling Trust maintains 'Top 50' lists of the heaviest fish of 14 species (barbel, [common] bream, [common] carp, chub, dace, eel, grayling, perch, pike, roach, rudd, tench, wels catfish and zander) but not of silver bream. The roll call of record and former record fish in this book can serve as an incomplete list partially filling this gap, albeit that large, verified fish caught beneath the weight of the record current at the time are still excluded.

SILVER BREAM MATCH CATCHES

Speaking with my match-angling friends, it would appear that none set out to target silver bream. Common bream present a weightier alternative when present. However, where they are present, silver bream do tend to come along as by-catch when anglers set out their stall for other mainstream match-angling targets such as common bream, roach, perch and tench, and are welcome make-weights on that basis.

SILVER BREAM AS BAIT

Pike and zander are widespread predators in British waters, consuming a broad range of other fish species (including their own kind).

Many is the time that I have had silver bream, along with roach, dace and others, taken off the hook when playing them. Almost anything sending out vibrations as it struggles in the water when retrieved can attract the attention of predatory fish – even large perch and chub – and silvery fish can also provide powerful visual stimulation. Silver bream score highly on both counts!

I am not a fan of live bait fishing. However, I will take some fish from waters where they are prolific and often stunted, and where I have secured permission from the fishery owner or manager. I freeze them after sealing them individually in vacuum seal bags promptly after killing them humanely. The vacuum seal bags ensure that the baits remain fresh and that they avoid getting 'freezer burned'. Frozen individually, only sufficient baits can be taken at any one time for a predator fishing session. My main predator sessions tend to be short, so I generally thaw them before use, but for a longer session they can be carried to the water in a cool box containing freezer blocks so that unused baits can be returned to the freezer unimpaired.

Silver bream hybrids are suitable, though I am sure I have accidentally also frozen a few small pure silvers as well. I have also then proceeded to catch both pike and zander on these baits after thawing (noting that a chilly bait may be rejected or ejected by a picky predator). I have also had big perch and brown trout pick up and get hooked on my pike dead baits.

Where pike are absent, a single hook on a heavy monofilament trace (17–20 lb breaking strain) can suffice. However, if there is a risk of a pike picking up the bait, a wire trace – as flexible and fine as you can find but always strong – is essential as it is potentially lethal to leave hooks, particularly treble hooks, in a lost fish. The reel line has to be proportionately strong too: better to fail to hook a suspicious fish than to leave the hooks in one.

SILVER BREAM AS PROBLEMS

In some waters that are favourable for silver bream and where there is little competition – I fondly recall one such Kentish farm pond back in the 1980s – silver bream can be prolific. Like rudd and perch, they can form dense, stunted populations in suitable small, enclosed waters.

In these situations, shoals of smaller silver bream can get to the bait before larger fish – roach, carp, tench, rudd or bigger silver bream – have a chance to home in on the angler's offering.

One of the ways to try to bypass these clouds of tiddlers is to put bait down directly to the pond bed, leaving none slowly sinking and attracting the unwanted attention of smaller fish. This can be achieved with a bait dropper, a swimfeeder or as part of a dense ball of groundbait that sinks rapidly before breaking up.

Four

My prior books dedicated to individual species have included roach (my favourite with two books!) as well as dace, gudgeon, ruffe and burbot, not to mention several other books addressing all British freshwater fish species collectively.

All in one way or another address the societal context of the species. But this is where the silver bream causes me more difficulty than many other fish. Relatively few cultural associations have been made or recorded with this most neglected of British freshwater fish species. In fact, some sources are quite dismissive of it. The authoritative Fishbase website says of the silver bream:

> Of little interest to game fishers and consumers. . . . Unpopular with commercial fishers due to its small size and competition with more desired species.

DOI: 10.1201/9781003311041-4

Here we go then, mining as much as we can of the social values associated with this charming and under-appreciated freshwater fish.

SILVER BREAM ETYMOLOGY: COMMON NAMES

The common name of the silver bream holds no deep mysteries. After all, it is silvery – very silvery in fact! – and also distinctly bream-like.

The name 'bream' is a derivation of the Middle English word 'breme', a name of Old French origin.

Eric Marshall-Hardy, then-editor of *Angling* magazine, wrote in his delightful little 1943 book *Coarse Fish*:

> The derivation of the French *Brême*, from which we get the word Bream or Breme, is not clear.

Another English common name applied to the silver bream is 'white bream', the silvery colour of the fish so bright as to appear white. The names 'bream flat' and 'breamflat' are equally descriptive of its laterally flattened shape.

Another common name is the 'Pomeranian bream', a name that we will explore in more detail later in this chapter.

SILVER BREAM ETYMOLOGY: FOREIGN NAMES

Silver bream occur across a broad Eurasian range, spanning multiple linguistic regions. Consequently, this fish goes by an even wider range of common names in the languages, and the local regions, where it occurs. These names include:

Language	Common names
Bulgarian	Bellitza
Czech	Cejn malý, Cejnek malý, Křínek, Piest, Skalák
Danish	Flire
Dutch	Blei, Kolblei
English	Silver bream, White bream, Flat bream, Bream Flat

Language	Common names
Estonian	Nurg
Finnish	Pasuri
French	Blike, Brème bordelière, Brème bordelière, Brémette, Hazelin, Petite brème
German	Blätte, Bleiche, Bleinse, Blenke, Bleyzer, Blicke, Blicke, Blicke, Bliengge, Breitfisch, Bresen, Bunke, Fliengge, Geiserze, Gieben, Giester, Grasblecke, Grastaschel, Güster, Güster, Halbbrachsen, Halbbressen, Hebber, Hörsel, Jüster, Kleinbreiser, Kulbauge, Mackel, Meckel, Plattfisch, Pleinzen, Pletten, Pliete, Plieten, Prünke, Rotflosser, Rotplieten, Sandblecke, Scheiber, Schniber, Steinmappen, Zobelpleinzen
Hungarian	Karika keszeg
Italian	Blicca
Latvian	Plicis
Lithuanian	Plakis
Norwegian	Flire, Flisebrasme
Spanish	Albur, Brema blanca
Persian	Seamparak
Polish	Kiełb, Kiełb pospolity, Krap
Rumanian	Batca, Batcă, Cârjanca, Cosac
Russian	Пескарь (Peskar), Gustera, Ploskyrka, Густера
Serbian	Krupatica
Slovak	Piest zelenkavý
Slovene	Androga
Swedish	Björkna
Turkish	Abdalca, Çapak, Tahta baligi
Welsh	Merfog gwyn

SILVER BREAM ETYMOLOGY: LATIN NAMES

Whilst the scientific name *Blicca bjoerkna* (Linnaeus 1758) has long and generally been accepted in science, breaking this name down into its component parts tells us something about this fish.

In his 1879 book British Fresh-Water Fishes, The Reverend W. Houghton notes:

> The specific name, blicca, is from the Anglo-Saxon verb blican, 'to shine', to which also the name Bleak (Alburnus lucidus), another Cyprinoid, must be referred.

As regards the species name bjoerkna, we revert to Swedish origins. Carl Linnaeus, the Swedish naturalist and physician who formalised the Latinate binomial nomenclature used to classify species, known after his ennoblement as Carl von Linné, was the person bestowing this specific name on the fish. In Swedish, bjoerk (Björk, Björck, Biörck or Bjork) means 'birch', and more specifically the common downy birch tree (Betula pubescens), which is also known as the moor birch, white birch, European white birch or hairy birch. The downy birch is closely related to, and frequently confused with, the silver birch (Betula pendula), sharing the familiar silver-white bark. This silver-white coloration appears to have been applied to the silveriness of the silver bream, the Swedish name for which is Björkna.

The extension to the Latin name is known as the 'authority', naming the person who first described it scientifically and the date it was described. The extension to the Latin name of the silver bream (Linnaeus, 1758) tells us that Carl Linnaeus was the first to describe it. The date 1758 is when it was first described.

However, the parentheses (brackets) around the author's name and date of description denote that the species was originally described under a different name. The original Latin name that Carl Linnaeus bestowed on the fish was Cyprinus bjoerkna Linnaeus, 1758; Linnaeus initially classified a lot of carp-like fishes under the wide genus Cyprinus.

The parentheses also alert readers that synonyms exist for the species. These synonyms are in fact diverse, many listed in the following tables. The first table includes synonyms that were classifications of the same fish as a different species, subsequently reclassified, for which the authority is not in brackets. These include, for example, *Cyprinus plestya* Leske, 1774. The second table lists synonyms of alternative names formerly applied to the same fish, such as *Abramis bjoerkna* (Linnaeus, 1758).

Latin names for formerly identified species subsequently identified as silver bream

Cyprinus gieben Wulff, 1765
Cyprinus plestya Leske, 1774
Cyprinus blicca Bloch, 1782
Cyprinus latus Gmelin, 1789
Cyprinus Meckel Hermann, 1804
Cyprinus laskyr Güldenstädt, 1814
Cyprinus gibbosus Pallas, 1814
Blicca argyroleuca Heckel, 1843
Abramis erythropterus Valenciennes, 1844
Abramis micropteryx Valenciennes, 1844
Cyprinus latus Gronow, 1854
Blicca intermedia Fatio, 1882
Blicca bjoerkna transcaucasica Berg, 1916
Blicca bjoerkna derjavini Dadikyan, 1970

Synonyms previously applied to silver bream

Cyprinus bjoerkna Linnaeus, 1758
Abramis bjoerkna (Linnaeus, 1758)
Blicca bjoerkna bjoerkna (Linnaeus, 1758)
Abramis björkna (Linnaeus, 1758)
Blicca bjoekna (Linnaeus, 1758)
Blicca bjorkna (Linnaeus, 1758)
Blicca björkna (Linnaeus, 1758)
Abramis blicca (Bloch, 1782)

Izaak Walton conspicuously did not write about the silver bream, though bream (common bream) were devoted a chapter in his 1653 *The Compleat Angler*. Walton was most likely not aware that some of the smaller bream with which he was familiar were in reality a different species.

THE POMERANIAN BREAM

The German ichthyologist Marcus Elieser Bloch, in his 1782 book *Oeconomische Naturgeschichte der Fische Deutschlands. Erster Theil*, was the first to describe an apparently new species of bream: the Pomeranian Bream. Bloch gave this new 'species' the scientific name *Cyprinus Buggenhagii*, deriving the specific name from a gentleman named M. Buggenhagen from whom he had received specimens. These specimens had been collected from Swedish Pomerania, a historical region on the southern shore of the Baltic Sea in Central Europe spanning current-day Poland and Germany. It is this region from which the common name derives. Various subsequent ichthyologists have changed the scientific name, offering similar, modified or conflicting descriptions.

In his 1866 book *British Fishes, Volume 2*, Robert Hamilton reported of the Pomeranian bream:

> Its introduction into the British Fauna we owe to Mr. Yarrell, who obtained from Mr. Brandon a fine specimen, captured in a net at Dagenham Breach.

William Yarrell had listed the Pomeranian bream in his 1859 book *A History of British Fishes in Two Volumes* under the Latin name *Abramis Buggenhagii*, keeping Bloch's specific name but transferring the fish into the genus *Abramis*. Today, only one species is in the genus *Abramis* (the common bream, *Abramis brama*), though other fishes had formerly been placed into that genus, including one of the former names given to the silver bream (see prior discussion of the now abandoned name *Abramis bjoerkna*).

In his 1879 book British Fresh-Water Fishes, The Reverend W. Houghton covers the Pomeranian bream in some detail, though clearly doubts its veracity as a distinct species, stating:

> *The fish to which Mr. Yarrell has given then name the Pomeranian Bream, from the country from which it was first discovered by Bloch, viz. Swedish Pomerania, and which has been generally regarded as a species of Abramis distinct from the Common Yellow and White Breams, is in all probability a hybrid between the Common Yellow Bream (Abramis brama), and the Roach (Leuciscus rutilus).*

Houghton continues:

> *The Pomeranian Bream has been admirably described by Dr. Günther as 'a Roach-like modification of the Bream, or a Bream-like modification of the Roach'.*

In his 1941 book The Observer's Book of Freshwater Fishes of the British Isles, A. Lawrence Wells wrote:

> *The so-called Pomeranian Bream, in which the lower fins are red, is probably a hybrid between this species and the roach or rudd.*

Given the somewhat variable descriptions of the Pomeranian bream, it is possible that some have described hybrids whilst others have described a silver bream. Whatever the reality of historical descriptions of specimens of uncertain heritage, today the Pomeranian bream is no longer seen as a discrete species but synonymous with the silver bream, *Blicca bjoerkna*.

SILVER BREAM CUISINE

The Reverend W. Houghton wrote in his 1879 book *British Fresh-Water Fishes* that the "White Bream, or Breamflat" (under the now-abandoned Latin name *Abramis blicca*)

. . . is worthless as an article of diet.

Also, Okan Yazıcıoğlu and colleagues in a 2017 scientific paper (see Bibliography) state:

The white bream does not have any commercial value because of its unpleasant taste . . . and having a great number of intermuscular bones.

These are hardly the most supportive endorsements of the gastronomic virtues of the silver bream! Little or nothing else specifically has been published relating to silver bream dining.

However, an internet search about silver bream as food was interesting. A British website specialising in Russian and European food products listed "Dried salted white bream ± 190g", of Russian origin. Tasty! Various other closely related dried freshwater fishes are available according to Russian websites. These include vobla. 'Vobla' in fact describes the salt-drying process used to preserve the fish, rather than relating to a specific fish species. Generally, vobla are Caspian roach (*Rutilus heckelii*), though it is likely that there is variability in the

Silver Bream and People

actual fish species used; the closely related silver bream may well end up thus prepared, as illustrated by the Russian 'Dried salted white bream' noted earlier.

One of the websites found on this search described vobla as

> . . . an ideal snack for beer. Has a pleasant aroma and does not need anything that could emphasize its taste. Dried vobla in vacuum packing is perfectly preserved, without requiring any culinary processing before consumption. The mild taste, as well as the characteristic spicy smell of this fish, will not leave indifferent the true fish gourmets.

Another is that

> You cannot eat dry salty Bream without beer – and plenty of it – preferably Russian beer.

A Russian saying is that

> If you are Russian and don't like Vobla and beer, then you are not Russian.

As noted previously, Izaak Walton did not refer to silver bream and seems to have been unaware of this distinct species. However, in the chapter 'The Fourth Day: Observations of the bream; and directions to catch him' in the 1653 *The Compleat Angler*, Walton wrote of bream:

> From St. James's-tide until Bartholomew-tide is the best; when they have had all the summer's food, they are the fattest.
> But though some do not, yet the French esteem this fish highly; and to that end have this proverb 'He that hath Breams in his pond, is able to bid his friend welcome'; and it is noted, that the best part of a Bream is his belly and head.

Also generically about 'bream', Eric Marshall-Hardy wrote in his 1943 book *Coarse Fish*:

> I believe that Jewish people also value Bream as food. For myself, I hold that the flesh is soft and flavourless, save of mud, which calls for savouries worthy of better flesh.

SILVER BREAM IN AQUACULTURE

Silver bream also have a lesser supporting rather than a lead role in aquaculture. Nick Giles writes of farming Wels catfish in his 1994 *Freshwater Fish of the British Isles: A Guide for Anglers and Naturalists*:

> Wels are a popular table fish on the Continent, with few bones and tasty, well-textured flesh; in Hungary they are cultured on fish farms where they are fed on white bream and other 'undesirable' species and marketed at weights of 3–4kg.

SILVER BREAM AND THE CREATIVE ARTS

Unlike Schubert's 'Die Forelle' (the trout) and Henry Williamson's 'Salar the Salmon', silver bream have not found their way into classic works of music and literature.

Likewise, silver bream seem not to have featured in heraldry. Nor have they become creatures inspiring myth and legend. In part, the relatively late distinction of silver bream as a true species may be part of this, though neither does their small stature help.

Silver bream have, however, been the focus of artistic inspiration, if not generally and with one main exception of high art. Many of the fish-focused books cited in this small volume carry illustrations of silver bream. Regrettably, few such illustrations are out of copyright, limiting my ability to use them in this book. However, one classic image is of a silver bream and a 'Pomeranian bream' together, painted by Alexander Francis Lydon (1836–1917). Lydon was a British watercolour artist, illustrator and engraver of natural history and landscapes. Lydon's painting is one of a set illustrating The Reverend W. Houghton's 1879 book British Fresh-Water Fishes and now widely used in a variety of tableware and other products.

SILVER BREAM AND NATURE CONSERVATION

Aquatic wildlife, including fish, is vitally important for the functioning and vitality of complex food webs in fresh waters. The vitality or demise of natural fish populations, as indeed all of the aquatic plants and animals with which they interact, should be a matter of conservation concern, for their inherent importance but also the many benefits that fresh-water ecosystems provide to humanity.

Freshwater ecosystems are particularly vulnerable as they integrate pressures across entire catchment landscapes. They are regarded as amongst the most threatened ecosystem types globally. Fishes are amongst the most vulnerable elements of the biodiversity of freshwater habitats. Thirty-eight per cent of Europe's freshwater fish species are threatened with extinction. A further 12 European fish species have already been declared extinct. Globally, 20% of the world's 10,000 freshwater fish species are listed as threatened, endangered or extinct.

Freshwater fish species, including, for example, Atlantic salmon, European eel and spined loach, are explicitly recognised in nature conservation legislation as threatened and requiring particular conservation attention. However, silver bream are not identified as a priority under various strands of nature conservation legislation.

Under the 'Red List' (IUCN 'Red List of Threatened Species'), documenting extinction threat, silver bream are listed as 'Least Concern' (LC).

The Bern Convention ('The Bern Convention on the Conservation of European Wildlife and Natural Habitats 1979'), the European Union (EU) Habitats Directive (Council Directive 92/43/EEC on the Conservation of Natural Habitats and of Wild Fauna and Flora) and UK nature conservation legislation (for example, the Wildlife and Countryside Act 1981, as subsequently amended) do not list silver bream for any special nature conservation attention. However, the Bern Convention and the EU Habitats Directive both impose bans on a number of listed destructive fishing methods.

PET SILVER BREAM

Silver bream are handsome fish, and many is the time I have thought about bringing a smaller specimen home for one of my aquaria. The trouble though is that they grow! And releasing captive fish – fed regularly, unprepared for detecting and evading predators, distanced from social interaction with other and cossetted by stable temperatures – is problematic as they may not fare well when released into the wild.

Some professional friends of mine have kept them captive, and I am sure that some might have found their way into large public aquaria. However, unless in large public exhibitions, this fish tends to grow too large for captivity in home aquaria.

Some aquarist stockists list silver bream as available for stocking as pond fish, noting their gregarious nature and that these fish are best kept in small shoals. They can reportedly be conditioned to accept food both on the bed of the pond and rising to take dried flake and pellet food.

Silver bream are, though, a handsome fish and perhaps could or should be better appreciated by fish-keepers as much as by anglers and the wider public.

THE ECONOMICS OF SILVER BREAM

Where present in mixed fish stocks, silver bream can make important contributions to harvestable natural resources of freshwater ecosystems. For example, they are cited in the 2013 Annual Report of the Danube Delta Biosphere Reserve (see Bibliography) under the heading 'Natural resources utilization, traditional activities':

> The main renewable natural resources used in 2013 were the fish and the reed. The fish capture recorded a value of around 1,312 tones and was dominated by fresh water species: Abramis brama, Carassius carassius, Rutilus rutilus, Aspius aspius, Carassius gibelio, Cyprinus carpio, Esox lucius, Perca fluviatilis, Silurus glanis, Stizostedion lucioperca and Blicca bjoerkna. In the 2012–2013 harvesting season, the quantity of reed harvested was of 3,019 tones.

In consideration of Pet silver bream, it is clear that the value of silver bream as pets is somewhat limited.

However, fish have worth beyond their utilitarian, harvestable and tradable values. Reflecting one element of non-utilitarian value, The Reverend W. Houghton in his 1879 book British Fresh-Water Fishes damns the silver bream with faint praise:

> . . . it is valued simply as affording food for Pike and other voracious fishes.

Bent J. Muus and Preben Dahlstrom were equally dismissive in their 1967 book *Collins Guide to the Freshwater Fishes of Britain and Europe*:

> The white bream is a competitor for food with the common bream, perch and eels.

Silver bream make some incremental contributions to the economics of angling, as part of pleasure and match catches and also as a specimen target for a niche market. To this we can add the contribution of silver bream to artistic inspiration.

However, the values of silver bream – or all fishes and wildlife – are far more diverse and deeper than that. Fish are integral elements of the ecosystems of which they are part, sometimes as prominent species and at others as minor constituents. Yet, in all of the ecosystems in which they occur, they play equally integral roles in the cycles of energy and nutrients and the wider integrity and functioning of the whole system. Representing this in narrow financial valuation terms is largely meaningless, as the vital significance of ecosystem functioning is as hugely important as it is unquantifiable. What is certain is that life for all is impoverished when ecosystems break down.

SILVER BREAM SOCIETIES

Many fish species have associated societies. If you are interested in pike or barbel, carp, tench or perch, roach and pretty much any other fish species, including tiddlers such as gudgeon and ruffe, you are well served by dedicated societies.

However, no such honour appears to have been extended to the silver bream at the time of writing.

There is, though, a private 'Silver Bream Angling Association' Facebook (social media) site at the time of writing, though, when I contacted the administrator in late 2021, it turned out that this was just a group for like-minded anglers rather than one focusing on silver bream as a target species.

Still, then, the much-neglected silver bream is in need of a species society of aficionados all of its own.

SILVER BREAM: NEGLECTED NO MORE?

The title of this book is *Silver bream: Britain's Most Neglected Freshwater Fish*. This is both a provocation and a former reality.

A new reality is that this much-overlooked fish now has a book dedicated entirely to it. The provocation is that we should all appreciate them just a little bit more!

Silver Bream Bibliography

The following works are referenced in this book, with my thanks to the authors concerned where quoted.

Baltic Marine Environment Protection Commission (Helsinki Commission). (2004). *Assessment of Coastal Fish in the Baltic Sea.* Baltic Sea Environment Proceedings No. 103A [Online].

Bell, Arthur P. (1926). *Fresh-Water Fishing for the Beginner.* Warne's Recreation Books, Frederick Warne & Co. Ltd., London.

Berg, C., Boogaart, J., Sibbing, F. and Osse, J. (1994). Implications of gill arch movements for filter feeding: an x-ray cinematographical study of filter-feeding white bream (*Blicca bjoerkna*) and common bream (*Abramis brama*). Journal of Experimental Biology, 191(1), pp. 257–282.

Bloch, Marcus Elieser. (1782). *Oeconomische Naturgeschichte der Fische Deutschlands: Erster Theil.* Hesse, Berlin.

Charman, Duncan. (2012). *Evolution of an Angler.* M Press (Media) Ltd., Maldon (Essex).

Danube Delta Biosphere Reserve Authority. (2013). *Danube Delta Biosphere Reserve Authority: Annual Report for 2013.* Ministry of Environment and Climate Change, Romania.

Everard, Mark. (2013). *Britain's Freshwater Fishes.* Princeton University Press/WildGUIDES, Princeton.

Everard, Mark. (2020). *The Complex Lives of British Freshwater Fishes.* CRC/Taylor and Francis, Boca Raton and London.

Giles, Nick. (1994). *Freshwater Fish of the British Isles: A Guide for Anglers and Naturalists.* Swan Hill, Shrewsbury.

Hamilton, Robert. (1866). *British Fishes, Volume 2.* Chatto & Windus, London.

Häunfling, B. and Brandl, R. (2000). Phylogenetics of European cyprinids: insights from allozymes. Journal of Fish Biology, 57(2), pp. 265–276.

Houghton, MA, FLS, The Reverend W. (1879). *British Fresh-Water Fishes.* William Mackenzie, London (Note: This book has been reprinted over the decades by numerous publishers, for example by The Peerage Press, London, in 1981).

Kottelat, M. and Freyhof, J. (2007). *Handbook of European Freshwater Fishes*. Publications Kottelat, Cornol and Freyhof, Berlin. pp. 646.

Matondo, B.N., Nlemvo, M., Ovidio, M., Poncin, P. and Philippart, J-C. (2008). Fertility in first-generation hybrids of roach, *Rutilus rutilus* (L.), and silver bream, *Blicca bjoerkna* (L.). *Journal of Applied Ichthyology*, 24(1), pp. 63–67.

Matondo, B.N., Ovidio, M., Philippart, J-C. and Poncin, P. (2012). A comparative study of sexual product quality in F1 hybrids of the bream *Abramis brama* × the silver bream *Blicca bjoerkna*. *Fisheries Science*, 78, pp. 1173–1178.

Marshall-Hardy, Eric. (1943). *Coarse Fish*. Herbert Jenkins, London.

Muus, Bent J. and Dahlstrom, Preben. (1967). *Collins Guide to the Freshwater Fishes of Britain and Europe*. Collins, London.

Newdick, J. (1983). *The Complete Freshwater Fishes of the British Isles*. A&C Black (Publishers) Ltd., London.

Pinder, A.C. (2001). *Keys to Larval and Juvenile Stages of Coarse Fishes from Fresh Waters in the British Isles*. Freshwater Biological Association Scientific Publications Volume 60. Freshwater Biological Association, Windermere.

Scott, N. (1969). *Ladybird Book about Coarse Fishing*. Ladybird Books Ltd, Loughborough.

Tombleson, Peter. (1954). *Bream: How to Catch Them*. Herbert Jenkins, London.

Tucker, D.E. (1987). *The Bristol Avon: Fish, Freshwater Life and Fishing*. Millstream Books, Bath.

Vetemaa, M., Kalda, R. and Tambets, M. (2008). Success of embryonic development of reciprocal hybrids of bream *Abramis brama* (L.) and white bream *Blicca bjoerkna* (L.). *Journal of Fish Biology*, 72(7), pp. 1787–1791.

Walton, Izaak and Cotton, Charles. (1653). *The Compleat Angler*. Maurice Clark, London (Available these days in many editions and from various publishers).

Wheeler, Alwyne. (1969). *The Fishes of the British Isles and North West Europe*. Michigan State University Press, East Lansing.

Wells. A. Lawrence. (1941). *The Observer's Book of Freshwater Fishes of the British Isles*. Frederick Warne and Co. Ltd., London.

Yarrell, William. (1859). *A History of British Fishes in Two Volumes*. John Van Voorst, London.

Yazıcıoğlu, O., Yılmaz, S., Yazıcı, R., Yılmaz, M. and Polat, N. (2017). Food items and feeding habits of white bream, *Blicca bjoerkna* (Linnaeus, 1758) inhabiting lake ladik (Samsun, Turkey). *Turkish Journal of Fisheries and Aquatic Sciences*, 17, pp. 371–378.

Yılmaz, S., Yazıcıoğlu, O., Yazıcı, R. and Polat, N. (2015). Age, growth and reproductive period of white bream, *Blicca bjoerkna* (l., 1758) in Lake Ladik, Turkey. *Journal of Limnology and Freshwater Fisheries Research*, 1(1), pp. 9–18.

Professor Mark Everard is a scientist, author and broadcaster, working on water and ecosystems around the world. He also has an irrationally large interest in fish!

Mark is a passionate angler, getting out whenever he can after coarse, game and sea fish. He has an enviable track record of specimen fish with a particular passion for roach, dace, mahseer and, of course, silver bream, and the whole river ecosystems that support them. He has long been a champion of the 'little fishes'.

Mark Everard is often referred to as Dr Redfin in the angling press for his special passion for roach.

For more on Mark and his work, see www.markeverard. co.uk.